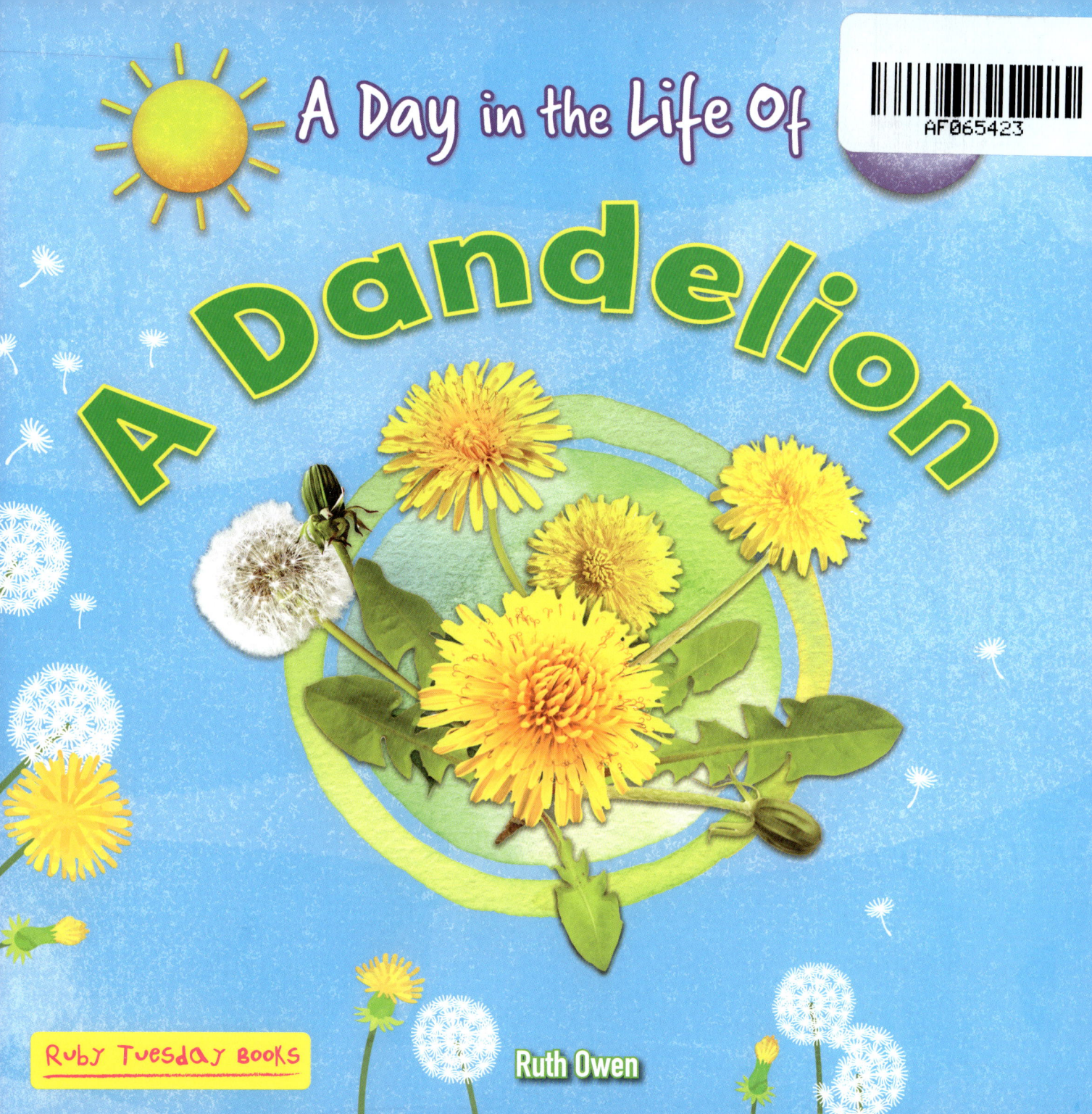

Published in 2025 by Ruby Tuesday Books Ltd.

Copyright © 2025 Ruby Tuesday Books Ltd.

All rights reserved. No part of this publication may be reproduced in whole or in part, stored in any retrieval system, or transmitted in any form or by any means, electronic, mechanical, photocopying, recording, or otherwise, without written permission from the publisher.

Editor: Mark J. Sachner
Design: Tammy West
Production: John Lingham

Photo Credits:
Alamy: 20 (Joe Blossom); Dreamstime: 9 (Ruslan Galiullin), 19 (Luayana), 21 (Kateryna Chyzhevska), 22T (Msundfors), 23T (Lucian Coman); iStockPhoto: 11; Nature Picture Library: 12 (Adrian Davies), 13 (Bence Mate); Ruby Tuesday Books: 4, 7, 8; Shutterstock: Cover (Quang Ho), 5 (O_Solara), 6 (LedyX), 10 (Lima_84), 14 (Bess Hamitii), 15 (Thilo Wagner), 16 (WCPW Photography), 17 (Serg Zastavkin), 18 (Petr Ganaj), 22C (Svetlana Zhukova), 22B (Midori Photography), 23C (irin-k), 23B (Lois GoBe), 24 (LedyX).

ISBN 978-1-78856-436-6

Printed in Poland by L&C Printing Group

www.rubytuesdaybooks.com

CONTENTS

Hello, Little Dandelion! 4

Glossary 22

Index 24

Hello, Little Dandelion!

It's a special day for a little dandelion plant.

Its fluffy **seedheads** wait for the wind to blow.

Today, the little dandelion will make new plants!

The little dandelion grows in the grass in a garden.

One week ago, the plant only had leaves.

Dandelion leaf

A dandelion is named for its **jagged** leaves.

Years ago, people thought they looked like a lion's teeth.

Soon, fat **buds** grew from the little plant.

Each bud opened and became a yellow flower.

Most people didn't notice the little dandelion.

But it was getting ready to do a very important job.

The dandelion's flowers made sweet **nectar** and dusty, yellow **pollen**.

Pollen

Bee

Nectar and pollen are food for bees and butterflies.

Munch Munch Munch

Grey squirrel

A dandelion's juicy flowers are food for many animals!

Some flowers did not get eaten.

They closed up and died, and seeds grew inside.

Flower **Closed-up flower** **Seedhead**

Then each closed-up flower opened and became a fluffy seedhead.

A dandelion's seeds are food, too!

puff puff puff

Now the wind blows the seeds from the seedheads.

The seeds float in the air on their fluffy **parachutes**.

Seed

Parachute

One by one

the seeds float to new growing places.

A raindrop PLOPS into a seed's parachute.

Raindrop

Snail

Just right for a tiny snail to take a drink!

Ant

Now, there's just one seed left.

An ant takes the seed to its nest for food.

This is a close-up picture of a dandelion seed.

The seed is tiny.

Dandelion seed

But inside the seed is everything that's needed to grow a new plant.

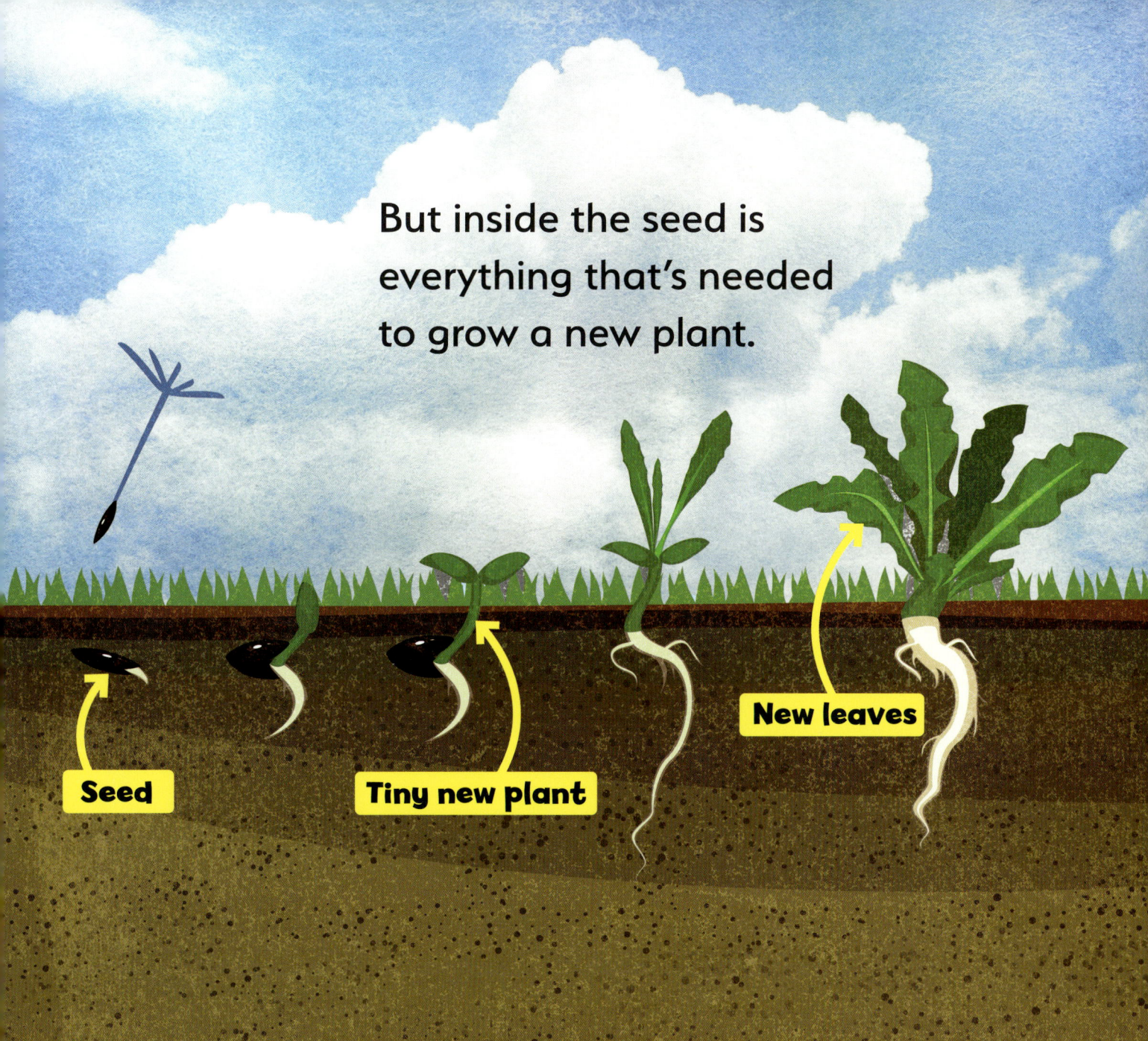

Now all that's left of the little dandelion plant are its leaves and stems.

Munch Munch Munch

The leaves become a juicy snack for a hungry pet tortoise.

The little dandelion's special day is over.

New dandelion plant

But soon its seeds will become
new plants!

Glossary

bud
A new part of a plant that opens to become a leaf or flower.

jagged
Having rough, pointed edges.

nectar
A sweet, sugary liquid that's made by flowers. It is food for bees and other insects.

parachute
A covering that fills with air and allows another object to slowly float to the ground.

pollen
A coloured dust that's made by flowers. It is food for insects and many other animals. Pollen helps flowers produce seeds.

Pollen

seedhead
A flower that has died and is now full of seeds.

Flower **Seedhead**

Index

A
animals 10–11, 13, 16–17, 20

B
buds 8

F
flowers 8, 10–11, 12

L
leaves 7, 19, 20

N
nectar 10

P
pollen 10

S
seedheads 4–5, 12–13, 14–15

seeds 4–5, 12–13, 14–15, 16–17, 18–19, 21